S0-AIL-945

THE LIBRARY OF PHYSICAL SCIENCE™

# Atoms and Chemical Reactions

Suzanne Slade

The Rosen Publishing Group's
PowerKids Press™
New York

*To Wesley and Charlie Pehling*

Published in 2007 by The Rosen Publishing Group, Inc.
29 East 21st Street, New York, NY 10010



First Edition

Editors: Melissa Acevedo and Amelie von Zumbusch
Book Design: Elana Davidian
Layout Design: Ginny Chu
Photo Researcher: Gabriel Caplan

Illustrations: pp. 8, 9 by Ginny Chu, adapted from an illustration by Tahara Anderson; pp. 12, 13 by Ginny Chu.

Photo Credits: Cover © Martyn F. Chillmaid/Photo Researchers, Inc.; p. 4 © Photodisc/Getty Images; p. 5 © Laguna Design/Photo Researchers, Inc.; p. 6 © Royalty-Free/Corbis; p. 7 © PhotoCuisine/Corbis; pp. 10, 18 (left) © Charles D. Winters/Photo Researchers, Inc.; p. 11 © Alfred Pasieka/Photo Researchers, Inc.; p. 14 © David Wrobel/Visuals Unlimited; p. 16 © Martyn F. Chillmaid/Photo Researchers, Inc.; p. 17 © Andrew McClenaghan/Photo Researchers, Inc.; p. 18 (right) © Andrew Lambert Photography / Photo Researchers, Inc.; p. 19 © Digital Stock; p. 20 © Michael Doolittle/The Image Works; p. 21 © Clouds Hill Imaging Ltd./Corbis.

Library of Congress Cataloging-in-Publication Data

Slade, Suzanne.
Atoms and Chemical Reactions / Suzanne Slade.— 1st ed.
p. cm. — (The library of physical science)
Includes index.
ISBN 1-4042-3415-2 (library binding) — ISBN 1-4042-2162-X (pbk.)
1. Atoms—Juvenile literature. 2. Chemical reactions—Juvenile literature. 3. Matter—Properties—Juvenile literature. I. Title. II. Series.
QC173.16.S525 2007
539.7—dc22
2005027851

Manufactured in the United States of America

# Contents

# Atoms and Molecules

Atoms are the basic building blocks of everything. A pencil, the Moon, and even your body are all made of atoms. Atoms are very small **particles**. They can only be seen with high-powered instruments that show very small things.

All atoms have smaller particles called protons, neutrons, and electrons in them. The **nucleus** in the center of an atom is made of protons and neutrons.

The red and white balls in this picture are protons and neutrons in the atom's nucleus. The yellow balls circling the nucleus are electrons.

This lactose molecule has three different types of atoms. Lactose is a kind of sugar found in milk.

Protons are positively charged and neutrons carry no charge. Tiny **negatively** charged electrons travel around, or orbit, the nucleus. The number of protons, neutrons, and electrons an atom has is what makes one type of atom different from another. Scientists have found more than 100 different kinds of atoms. When atoms connect with other atoms, they form **molecules**. Molecules combine with other molecules to create many types of matter.

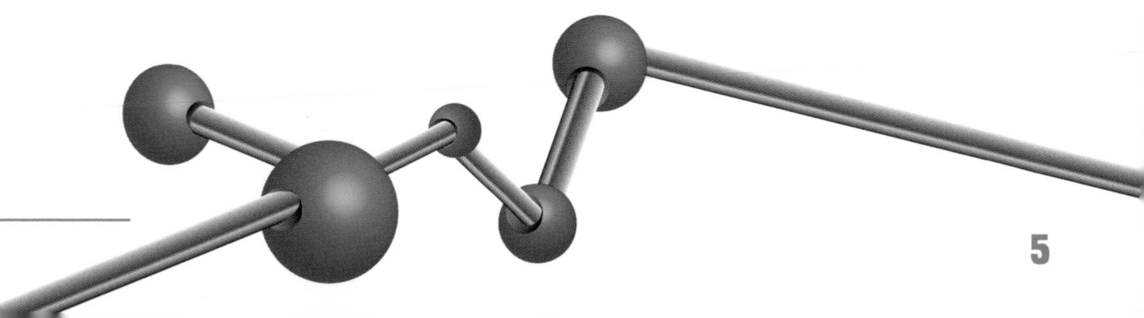

# Elements

If a **substance** is made of just one kind of atom, that substance is an element. There are 94 elements found in nature. Scientists have created 22 more elements in a lab. All elements have different properties, which include how they look, feel, or act under certain conditions.

This thermometer is filled with the liquid element mercury. Thermometers measure how hot something is.

An element can be a solid, liquid, or gas. Most elements are solid metals. Some metal

Helium is a gas element that is lighter than air. Therefore, a balloon full of helium floats.

elements you might find in your house are iron, silver, and gold. Common gaseous elements include helium, the gas that makes balloons float, and **oxygen**, which you breathe. Each element has a **symbol** of one, two, or three letters. The symbol for copper is Cu.

These pans are coated with the metal copper. Copper is one of the many elements that are metals.

Each element has a different atomic number. The atomic number is the number of protons in an atom of an element. The atomic mass of an element is the total number of protons and neutrons in an atom.

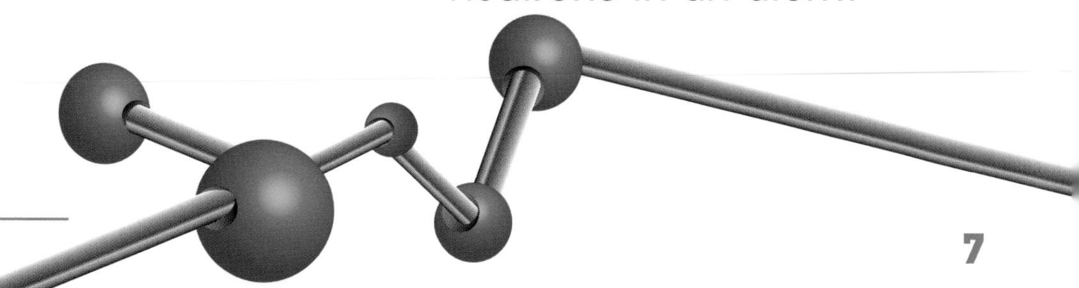

# The Periodic Table

All the known elements can be found on a chart called the periodic table. The periodic table lists the elements in the order of their atomic number. An element's position on the periodic table gives you facts about the element, such as its properties.

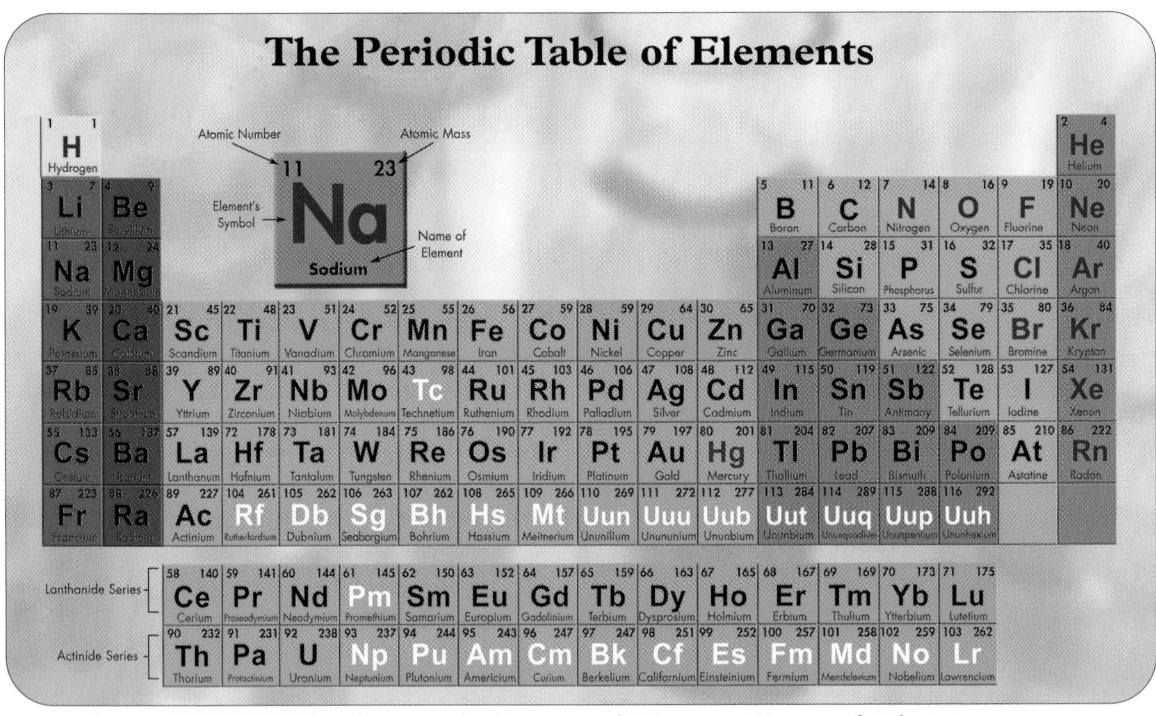


The periodic table, above, changes each time scientists find a new element. Dmitri Mendeleyev made the first periodic table in 1869.

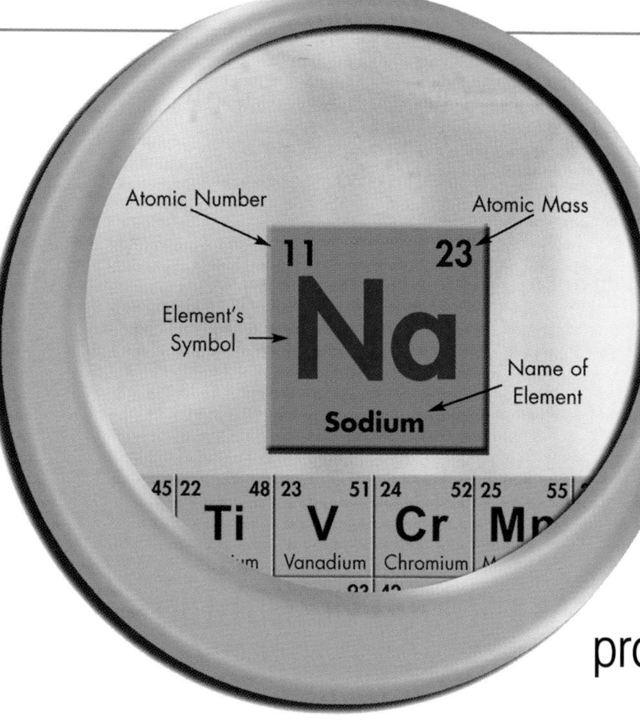


The symbol for the shiny metal sodium is Na. The symbol comes from *natrium* which is the Latin word for "sodium."

There are 18 **columns** on the periodic table. Each column is called a group. Elements in each group share many of the same properties. The rows of elements are called periods.

Every element has its own square on the periodic table. An element's name, symbol, atomic number, and atomic mass can be found inside this square. As scientists create new elements, they are added to the table. The periodic table has 116 elements so far.

# Compounds and Mixtures

When two or more elements join, they form a compound. A compound has different properties than do the elements used to create it. Once a compound is formed, it cannot come apart by itself.

Every compound has a molecular formula. The formula shows the symbols of the elements in the compound. The number of atoms of each element is also shown in a molecular formula. For example, the molecular formula for water is $H_2O$. H is the symbol for hydrogen. Hydrogen

This water molecule has two white hydrogen atoms and one red oxygen atom.

When you mix oil and water together their molecules do not join. Oil and water is a mixture. Its two parts can be separated again.

is a colorless gas that weighs less than any other element. The small two after the H means that a water molecule has two hydrogen atoms. Water also has one atom of oxygen. This is shown by the symbol O.

Some elements do not bond when they are mixed together. A mixture is a combination of two or more elements that can be separated and whose properties do not change. For example, a jar filled with sand and water is a mixture. You can separate the sand from the water.

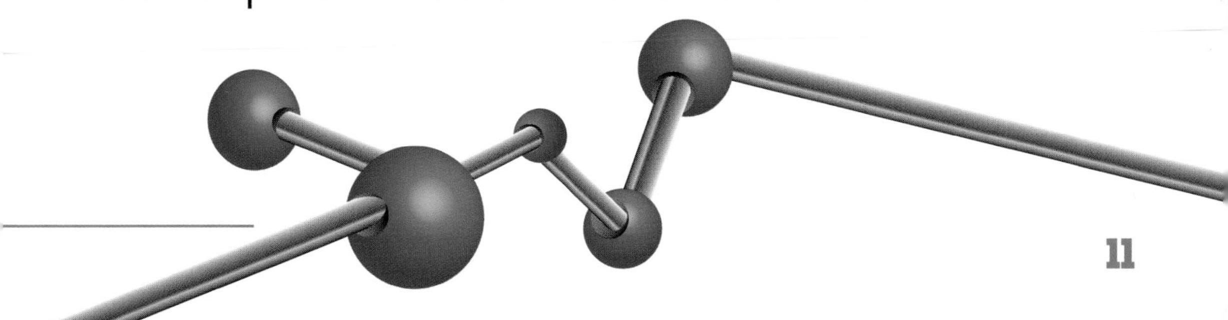

# Chemical Reactions

A chemical reaction occurs when matter mixes with other matter and causes a change. During such a reaction, two or more elements or compounds form a new product. The elements or compounds that change in a chemical reaction are called reactants.

The compounds that chemical reactions create are called products. A product has different properties from those of its reactants

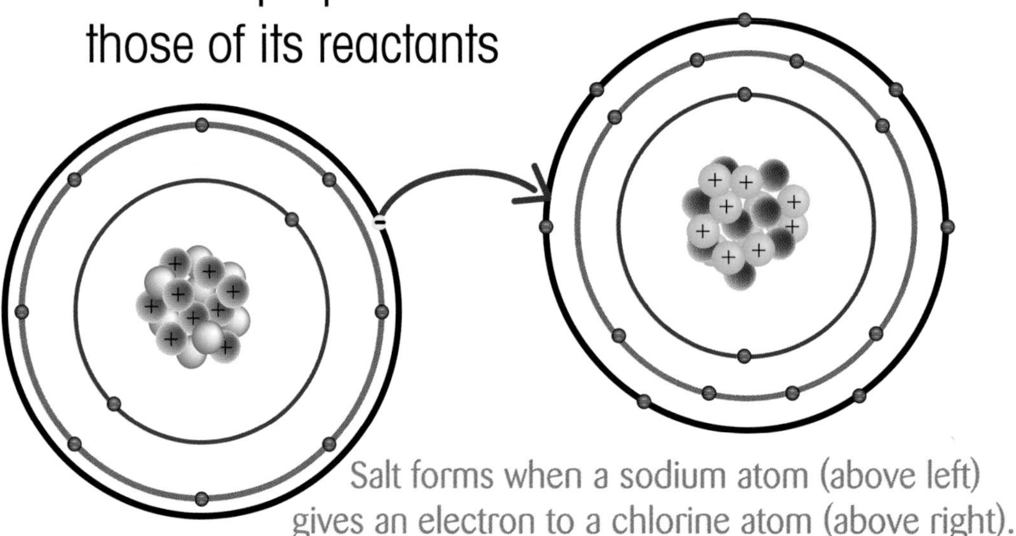

Salt forms when a sodium atom (above left) gives an electron to a chlorine atom (above right). This creates an ionic bond.

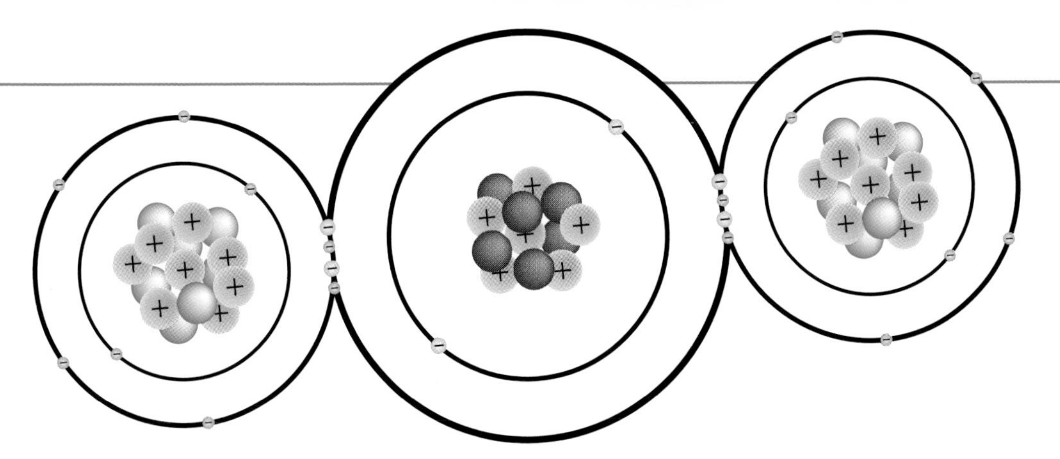


Carbon dioxide forms when a carbon atom shares its electrons with two oxygen atoms. The carbon atom is in the middle.

because it is a completely different substance. After a compound is formed, it cannot come apart by itself. Small, strong bonds hold the elements in a compound together. These bonds are created by the electrons in atoms.

An ionic bond is formed when one atom gives electrons to another atom. Table salt, shown by the formula NaCl, is a compound created by an ionic bond. Covalent bonds are formed when atoms share electrons. The gas **carbon dioxide** has two covalent bonds.

# Chemical Equations

A **chemical equation** shows how atoms rearrange in a chemical reaction. The atoms that start a chemical reaction are on the left side of a chemical equation. They are called reactants. A chemical reaction creates products. The products are shown on the right side of the equation.

An arrow between the reactants and products of a chemical equation shows the direction in which the atoms are changing. Usually this arrow goes from left to right. If a chemical equation has one

The bubbles in a glass of soda are carbon dioxide.

arrow that points in either direction, then that equation is reversible. In a reversible chemical reaction, products can combine with each other to create the reactants.

The number of atoms on both sides of a chemical equation always stays the same. Because atoms are not created or lost during a chemical reaction, the number of reactant atoms equals the number of atoms in the product. A molecule of carbon dioxide is made from one carbon atom and two oxygen atoms.

When you write the reactants for the chemical equation for this reaction, put a two in front of the O to show that you need two oxygen atoms.

# Recognizing New Compounds

In a chemical reaction, elements or compounds bond to form a new compound with different properties from those of its reactants. New compounds are recognized by their physical and chemical properties.

Physical properties are how a compound looks, feels, or acts. Chemical properties are how a compound changes when it is mixed with other compounds. One chemical property is a compound's **acid level**. Have you ever taken a bite of a lemon? Its sour taste is caused by

Scientists use special paper called litmus paper to discover a substance's acid level. A substance with acid, such as a lemon, turns the bluish litmus paper red.

**acid**. Many new compounds can be recognized by measuring their acid levels.

Other substances create solid compounds when they are mixed. For example, the liquid compounds potassium iodide and lead nitrate form a yellow solid called lead iodide when they are mixed. Another way to recognize new compounds is by color changes. For example, iodine, a substance used in drugs, turns blue when it touches starch. Starch is a substance found in food and drugs. Iodine is used to find starch compounds.

The starch in this test tube turns the iodine from orange to blue. Scientists use iodine to test if something has starch in it.

# Chemical Reactions and Salts

Do you like the taste of salt on french fries? Most people do. There are many other types of salt besides the kind that you eat. Salt compounds are used in some drugs. Farmers also use certain salts to help plants grow.

Sodium reacts so easily with other elements that it is almost never found on its own in nature.

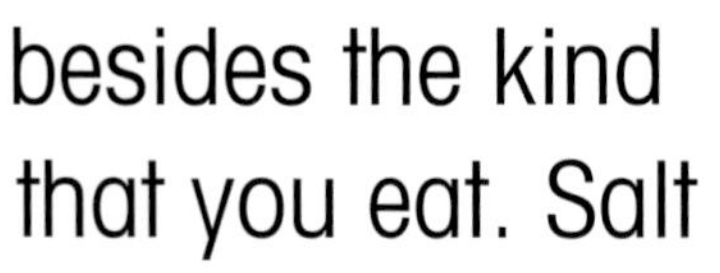

A salt is a compound that is created when a **base** joins an acid. For example, the shiny metal sodium combines with a

Chlorine is a yellowish green gas. It is used to clean water for drinking and in swimming pools.

gas called chlorine to make sodium chloride. Sodium chloride is a white salt like table salt. Though sodium is an unsafe metal and chlorine is a poisonous gas, they produce a harmless salt when they are mixed together.

There are many useful salts. People use the salt calcium chloride to melt the snow and ice on roads and sidewalks. Another type of salt, called potassium nitrate, is used to make fireworks.

The table salt you eat comes from seawater and from salt mines in the ground.

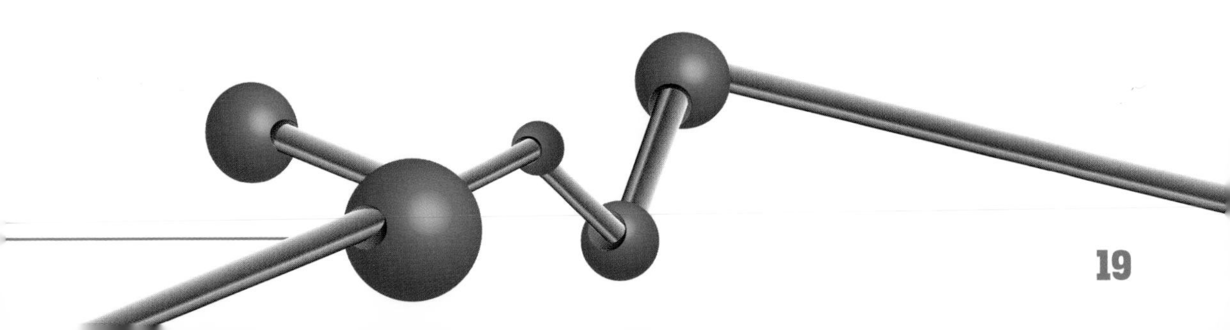

# Photosynthesis

**Photosynthesis** is a chemical reaction that happens mostly in the leaves of plants. Sunlight provides the power for this reaction. Photosynthesis also uses carbon dioxide from the air and water. Carbon dioxide and water are the reactants in this reaction. There

People depend on plants to make the oxygen we need to live.

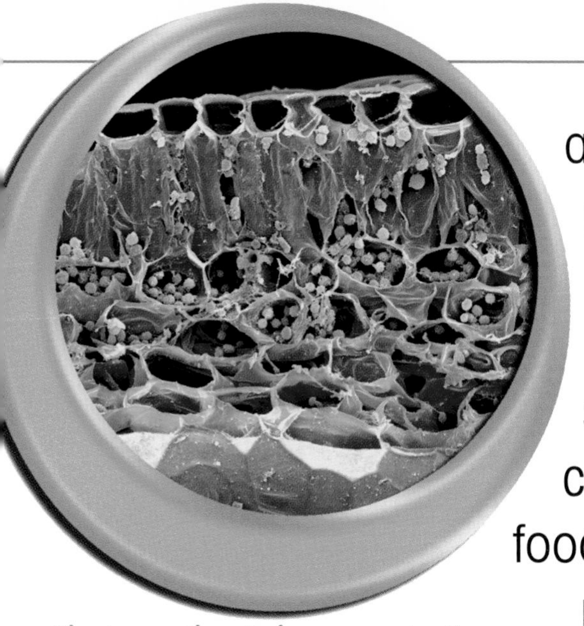

Photosynthesis happens in tiny parts of plants called chloroplasts. The chloroplasts in this close-up picture of a leaf are yellow.

are two products of photosynthesis. These products are sugar and oxygen. Plants use the sugar created in this chemical reaction for food. People and animals need the oxygen made in photosynthesis to breathe.

**Photosynthesis Reaction**

| $6CO_2$ | + | $6H_2O$ | ⇨ | $C_6H_{12}O_6$ | + | $6O_2$ |
|---|---|---|---|---|---|---|
| **carbon dioxide** | | **water** | | **sugar** | | **oxygen** |

Where do plants get the carbon dioxide they need for photosynthesis? They get it from you! People breathe in oxygen and breathe out carbon dioxide. People and plants depend on each other.

# Everyday Chemical Reactions

Chemical reactions are happening around you all the time. Your stomach uses chemical reactions to break down food. Your stomach has hydrochloric acid. Its chemical formula is HCl. This acid breaks down food so your body can take it in. Hydrochloric acid is very powerful. It can even break down metals!

Bread dough, or batter, rises because of a chemical reaction. The yeast in the dough turns sugar into carbon dioxide gas. The carbon dioxide bubbles make the dough rise. This makes the bread light when it is baked.

Some chemical reactions can be harmful. Old milk becomes sour, and metal rusts because of chemical reactions. Although not all chemical reactions are helpful, they are still a necessary part of your world.

# Glossary

**acid** (A-sid) Matter that breaks down matter faster than water does.

**acid level** (A-sid LEH-vul) The measure of matter's acidity.

**base** (BAYS) Matter that tastes bitter and joins with acids to form salts.

**carbon dioxide** (KAR-bin dy-OK-syd) A gas that plants take in from the air and use to make food.

**chemical equation** (KEH-mih-kul ih-KWAY-zhun) A statement that shows the products and reactants in a chemical reaction.

**columns** (KAH-lumz) Rows that go up and down.

**formula** (FOR-myuh-luh) A group of symbols and numbers that show what is in a compound.

**molecules** (MAH-lih-kyoolz) Two or more atoms joined together.

**negatively** (NEH-guh-tiv-lee) The opposite of positively.

**nucleus** (NOO-klee-us) Protons and neutrons joined together in the center of an atom.

**oxygen** (OK-sih-jen) A gas that has no color, taste, or odor and is necessary for people and animals to breathe.

**particles** (PAR-tih-kulz) Small pieces of something.

**photosynthesis** (foh-toh-SIN-thuh-sus) The way in which green plants make their own food from sunlight, water, and carbon dioxide.

**substance** (SUB-stans) Any matter that takes up space.

**symbol** (SIM-bul) The letter or letters that stand for an element.

# Index



# Web Sites

Due to the changing nature of Internet links, PowerKids Press has developed an online list of Web sites related to the subject of this book. This site is updated regularly. Please use this link to access the list:
www.powerkidslinks.com/lops/chemreac/